NOTE

Thank You for Choosing the Math Worksheet Bundle!
We have a wide range of educational products designed to
make teaching and learning fun and effective.
Here's how you can stay connected and continue to benefit from our materials:

✸ **Make a Purchase:** Explore our store and discover more
exciting educational resources that can enhance
your teaching experience.

✸ **Leave a Review:** If you've enjoyed our worksheets,
we'd greatly appreciate your positive feedback.
Your reviews help other educators like you find quality
resources.

Thank you once again for choosing our worksheets.
Warm regards,
Nisha

WORKSHEET 1
COUNTING BY 1S

Count for the following and write answers

Name: _____ Class: _____

1. Count from 1 to 20 aloud: ☐
2. Count the number of fingers and toes you have: ☐
3. Count how many students are in your classroom: ☐
4. Count the number of apples in a basket: ☐
5. Count the number of books on your bookshelf: ☐
6. Count the number of fingers on your two hands together: ☐
7. Count the number of letters in your first name: ☐
8. Count the number of windows in your house: ☐
9. Count the number of shoes in your closet: ☐
10. Count the number of buttons on your shirt or jacket: ☐

WORKSHEET 2
COUNTING BY 1S

Count for the following and write answers

Name: _____ Class: _____

1. Count the number of pencils in your pencil case. ☐
2. Count how many fingers and toes you have on both hands and feet combined. ☐
3. Count the number of chairs in your kitchen or dining area. ☐
4. Count the number of windows in your home. ☐
5. Count the number of fingers on one hand, then the other hand, and add the totals together. ☐
6. Count the number of candies in a bag or jar. ☐
7. Count the number of cars passing by on the street. ☐
8. Count the number of steps you climb to reach your front door. ☐
9. Count the number of pets you have, such as cats, dogs, or fish. ☐
10. Count the number of buttons on your remote control. ☐

WORKSHEET 3
COUNTING BY 1S

Count for the following and write answers

Name: _____ Class: _____

1. Count the number of leaves on a tree outside. ☐
2. Count the number of teeth you have. ☐
3. Count the number of chairs in your classroom. ☐
4. Count the number of fingers on both hands and toes on both feet together. ☐
5. Count the number of houses on your street. ☐
6. Count the number of petals on a flower. ☐
7. Count the number of tiles on your bathroom floor. ☐
8. Count the number of markers in a box. ☐
9. Count the number of people waiting in line at the grocery store. ☐
10. Count the number of stairs in your home. ☐

WORKSHEET 4
COUNTING BY 2S

Count for the following and write answers

Name: _____ Class: _____

1. Count the number of hands in your family (2 for each person). ☐

2. Count the number of legs on insects or spiders you see outside. ☐

3. Count the number of wheels on toy cars. ☐

4. Count the number of eyes on your favorite stuffed animals. ☐

5. Count the number of gloves you have (2 for each pair). ☐

6. Count the number of wings on a butterfly (2, 4, 6...). ☐

7. Count the number of legs on a cat (2, 4, 6...). ☐

8. Count the number of slices in a pizza (2, 4, 6...). ☐

9. Count the number of ears on a bunny (2, 4, 6...). ☐

10. Count the number of wheels on a skateboard (2, 4...). ☐

WORKSHEET 5
COUNTING BY 2S

Write the answers for the following:

Name: _____ Class: _____

1. Count the number of tails on a pair of kittens (2, 4, 6...): ☐

2. Count the number of wings on a dragonfly (2, 4, 6...): ☐

3. Count the number of laces on your shoes (2, 4...): ☐

4. Count the number of arms on a person (2, 4...): ☐

5. Count the number of pockets on your jacket (2, 4...): ☐

6. Count the number of wheels on a tricycle (2, 4...): ☐

7. Count the number of legs on a spider (2, 4, 6...): ☐

8. Count the number of books on a shelf, pairing them up (2, 4, 6...): ☐

9. Count the number of laces on your shoes (2, 4...): ☐

10. Count the number of ears on a pair of rabbits (2, 4...): ☐

WORKSHEET 6
READING AND WRITING NUMBERS UP TO 1,000

Name: _____ Class: _____

1. Read the number 836 as "eight hundred thirty-six." _____

2. Write the number 572 as "five hundred seventy-two." _____

3. Read the number 287 as "two hundred eighty-seven." _____

4. Write the number 964 as "nine hundred sixty-four." _____

5. Read the number 125 as "one hundred twenty-five." _____

6. Write the number 719 as "seven hundred nineteen." _____

7. Read the number 543 as "five hundred forty-three." _____

8. Write the number 368 as "three hundred sixty-eight." _____

9. Read the number 652 as "six hundred fifty-two." _____

10. Read the number 287 as "two hundred eighty-seven." _____

WORKSHEET 7
READING AND WRITING NUMBERS UP TO 1,000

Name: _____ Class: _____

1. Read the number 781 as "seven hundred eighty-one." _____

2. Write the number 295 as "two hundred ninety-five." _____

3. Read the number 543 as "five hundred forty-three." _____

4. Write the number 827 as "eight hundred twenty-seven." _____

5. Read the number 134 as "one hundred thirty-four." _____

6. Write the number 576 as "five hundred seventy-six." _____

7. Read the number 983 as "nine hundred eighty-three." _____

8. Write the number 246 as "two hundred forty-six." _____

9. Read the number 713 as "seven hundred thirteen." _____

10. Write the number 419 as "four hundred nineteen." _____

WORKSHEET 8

READING AND WRITING NUMBERS UP TO 1,000

Name: _____ Class: _____

1. Read the number 392 as "three hundred ninety-two." _____

2. Write the number 845 as "eight hundred forty-five." _____

3. Read the number 561 as "five hundred sixty-one." _____

4. Write the number 127 as "one hundred twenty-seven." _____

5. Write the number 914 as "nine hundred fourteen." _____

6. Read the number 378 as "three hundred seventy-eight." _____

7. Write the number 526 as "five hundred twenty-six." _____

8. Read the number 642 as "six hundred forty-two." _____

9. Write the number 739 as "seven hundred thirty-nine." _____

10. Write the number 658 as "six hundred fifty-eight." _____

WORKSHEET 9
UNDERSTANDING PLACE VALUE:

Name: _____ Class: _____

1. Identify the digit in the tens place in the number 672: ☐

2. Identify the digit in the tens place in the number 195: ☐

3. Identify the digit in the tens place in the number 364: ☐

4. Identify the digit in the tens place in the number 529: ☐

5. Identify the digit in the tens place in the number 821: ☐

6. Identify the digit in the tens place in the number 147: ☐

7. Identify the digit in the tens place in the number 356: ☐

8. Identify the digit in the tens place in the number 748: ☐

9. Identify the digit in the tens place in the number 613: ☐

10. Identify the digit in the tens place in the number 438: ☐

WORKSHEET 10
UNDERSTANDING PLACE VALUE:

Name: _____ Class: _____

1. Determine the digit in the hundreds place in the number 583: ☐

2. Determine the digit in the hundreds place in the number 492: ☐

3. Determine the digit in the hundreds place in the number 915: ☐

4. Determine the digit in the hundreds place in the number 357: ☐

5. Determine the digit in the hundreds place in the number 842: ☐

6. Determine the digit in the hundreds place in the number 631: ☐

7. Determine the digit in the hundreds place in the number 479: ☐

8. Determine the digit in the hundreds place in the number 268: ☐

9. Determine the digit in the hundreds place in the number 746: ☐

10. Determine the digit in the hundreds place in the number 726: ☐

WORKSHEET 11
UNDERSTANDING PLACE VALUE:

Name: _____ Class: _____

1. Explain the value of the digit 9 in the number 964 (it's in the tens place). | **900**

2. Explain the value of the digit 2 in the number 832 (it's in the tens place). | ☐

3. Explain the value of the digit 7 in the number 727 (it's in the tens place). | ☐

4. Explain the value of the digit 3 in the number 349 (it's in the tens place). | ☐

5. Explain the value of the digit 1 in the number 196 (it's in the tens place). | ☐

6. Explain the value of the digit 6 in the number 642 (it's in the tens place). | ☐

7. Explain the value of the digit 8 in the number 839 (it's in the tens place). | ☐

8. Explain the value of the digit 4 in the number 467 (it's in the tens place). | ☐

9. Explain the value of the digit 0 in the number 502 (it's in the tens place). | ☐

10. Explain the value of the digit 5 in the number 578 (it's in the tens place). | ☐

WORKSHEET 12
UNDERSTANDING PLACE VALUE:

Name: _____ Class: _____

1. **Identify the digit in the hundreds place in the number 567.** ☐

2. **Identify the digit in the hundreds place in the number 412.** ☐

3. **Identify the digit in the hundreds place in the number 846.** ☐

4. **Identify the digit in the hundreds place in the number 125.** ☐

5. **Identify the digit in the hundreds place in the number 749.** ☐

6. **Identify the digit in the hundreds place in the number 683.** ☐

7. **Identify the digit in the hundreds place in the number 271.** ☐

8. **Identify the digit in the hundreds place in the number 538.** ☐

9. **Identify the digit in the hundreds place in the number 365.** ☐

10. **Identify the digit in the hundreds place in the number 932.** ☐

WORKSHEET 13
UNDERSTANDING PLACE VALUE:

Name: _____ Class: _____

1. Write a number where the digit 2 is in the hundreds place: `267`

2. Write a number where the digit 2 is in the hundreds place: _____

3. Write a number where the digit 2 is in the hundreds place: _____

4. Write a number where the digit 2 is in the hundreds place: _____

5. Write a number where the digit 2 is in the hundreds place: _____

6. Write a number where the digit 2 is in the hundreds place: _____

7. Write a number where the digit 2 is in the hundreds place: _____

8. Write a number where the digit 2 is in the hundreds place: _____

9. Write a number where the digit 2 is in the hundreds place: _____

10. Write a number where the digit 2 is in the hundreds place: _____

WORKSHEET 14

ADD THE FOLLOWING NUMBERS

Name: _____ Class: _____

1. Subtract 7 from 15: _____

2. Add 9 and 5: _____

3. Subtract 4 from 11: _____

4. If you have 12 candies and you eat 3, how many candies do you have left? _____

5. What is 17 ÷ 2? _____

6. Subtract 8 from 14: _____

7. Add 7 and 8: _____

8. Subtract 5 from 9: _____

9. If you have 10 stickers and your friend gives you 4 more, how many stickers do you have now? _____

10. If you have 10 stickers and your friend gives you 4 more, how many stickers do you have now? _____

WORKSHEET 15
ADDITION AND SUBTRACTION

Name: _____ Class: _____

1. **Subtract 36 from 59.**

2. **Add 56 and 29.**

3. **Subtract 48 from 73.**

4. **Add 67 and 48.**

5. **Subtract 27 from 53.**

6. **Add 35 and 68.**

7. **Subtract 59 from 82.**

8. **Add 74 and 56.**

9. **Subtract 38 from 64.**

10. **Add 28 and 47.**

WORKSHEET 16
WORD PROBLEMS

Name: _____ Class: _____

1. You have 24 marbles. You give 9 to your friend. How many marbles do you have left?

2. There are 16 students in the classroom. 7 students are girls, and the rest are boys. How many boys are there?

3. There are 38 candies in one jar and 27 candies in another. How many candies are there in total?

4. Timmy had 14 toy cars. He got 9 more as a gift. How many toy cars does he have now?

5. There are 53 pencils, and 28 of them are blue. How many are not blue?

6. Sarah had 42 stickers. She gave 15 to her brother. How many stickers does she have now?

7. There are 65 books on the shelf, and 27 are fiction books. How many are non-fiction books?

8. Mary has 56 stickers, and Jane has 38 stickers. How many stickers do they have together?

9. There are 46 students on the playground. 18 of them are playing soccer, and the rest are playing basketball. How many are playing basketball?

10. There are 15 apples, and 8 are red. How many are not red?

WORKSHEET 17
ADDITION AND SUBTRACTION

Name: _____ Class: _____

1. Subtract 5 from 19:

2. Add 8 and 7:

3. Subtract 6 from 14:

4. If you have 11 balloons and you give away 3, how many balloons do you have left?

5. What is 16 ÷ 3?

6. Subtract 9 from 20:

7. Add 6 and 9:

8. Subtract 7 from 11:

9. If you have 15 candies, and your friend gives you 6 more, how many candies do you have now?

10. What is 12 ÷ 4?

WORKSHEET 18
ADDITION AND SUBTRACTION

Name: _____ Class: _____

1. Subtract 42 from 79: ☐
2. Add 65 and 38: ☐
3. Subtract 56 from 82: ☐
4. Add 79 and 46: ☐
5. Subtract 35 from 61: ☐
6. Add 57 and 68: ☐
7. Subtract 49 from 83: ☐
8. Add 78 and 57: ☐
9. Subtract 43 from 69: ☐
10. Add 37 and 58: ☐

WORKSHEET 19
SOLVING WORD PROBLEMS INVOLVING ADDITION AND SUBTRACTION

Name: _____ Class: _____

1	You have 28 marbles. You give 12 to your friend. How many marbles do you have left?	
2	There are 26 students in the classroom. 9 students are girls, and the rest are boys. How many boys are there?	
3	There are 54 candies in one jar and 28 candies in another. How many candies are there in total?	
4	Emily had 16 stickers. She gave 8 to her brother. How many stickers does she have now?	
5	There are 47 pencils, and 19 of them are blue. How many are not blue?	
6	Mark had 63 toy cars, and he bought 28 more. How many toy cars does Mark have now?	
7	There are 68 books on the shelf, and 32 are fiction books. How many are non-fiction books?	
8	Amy has 53 marbles, and Jane has 39 marbles. How many marbles do they have together?	
9	There are 57 students in the school gym. 25 students are playing basketball, and the rest are playing soccer. How many are playing soccer?	
10	There are 18 crayons, and 7 are red. How many are not red?	

WORKSHEET 20
COMPARISON OF NUMBERS: USE THE APPROPRITATE SYMBOL < OR > OR =

Name: _____   Class: _____

1) 7 ☐ 4
2) 6 ☐ 9
3) 5 ☐ 5
4) 12 ☐ 18
5) 3 ☐ 5
6) 8 ☐ 8
7) 25 ☐ 30
8) 10 ☐ 15
9) 7 ☐ 7
10) 21 ☐ 21

ANSWER KEYS

WORKSHEET 9

UNDERSTANDING PLACE VALUE (ANSWERS)

Name: _____ Class: _____

1. Identify the digit in the tens place in the number 438. The digit in the tens place is 3.

2. Identify the digit in the tens place in the number 672. The digit in the tens place is 7.

3. Identify the digit in the tens place in the number 195. The digit in the tens place is 9.

4. Identify the digit in the tens place in the number 364. The digit in the tens place is 6.

5. Identify the digit in the tens place in the number 529. The digit in the tens place is 2.

6. Identify the digit in the tens place in the number 821. The digit in the tens place is 2.

7. Identify the digit in the tens place in the number 147. The digit in the tens place is 4.

8. Identify the digit in the tens place in the number 356. The digit in the tens place is 5.

9. Identify the digit in the tens place in the number 748. The digit in the tens place is 4.

10. Identify the digit in the tens place in the number 613. The digit in the tens place is 1.

WORKSHEET 10

UNDERSTANDING PLACE VALUE (ANSWERS)

Name: _____ Class: _____

1. **Determine the digit in the hundreds place in the number 726. The digit in the hundreds place is 7.**

2. **Determine the digit in the hundreds place in the number 583. The digit in the hundreds place is 5.**

3. **Determine the digit in the hundreds place in the number 492. The digit in the hundreds place is 4.**

4. **Determine the digit in the hundreds place in the number 915. The digit in the hundreds place is 9.**

5. **Determine the digit in the hundreds place in the number 357. The digit in the hundreds place is 3.**

6. **Determine the digit in the hundreds place in the number 842. The digit in the hundreds place is 8.**

7. **Determine the digit in the hundreds place in the number 631. The digit in the hundreds place is 6.**

8. **Determine the digit in the hundreds place in the number 479. The digit in the hundreds place is 4.**

9. **Determine the digit in the hundreds place in the number 268. The digit in the hundreds place is 2.**

10. **Determine the digit in the hundreds place in the number 746. The digit in the hundreds place is 7.**

WORKSHEET 11

UNDERSTANDING PLACE VALUE (ANSWERS)

Name: _____ Class: _____

1. Explain the value of the digit 5 in the number 578 (it's in the tens place). The digit 5 in the tens place represents 50.
2. Explain the value of the digit 9 in the number 964 (it's in the tens place). The digit 9 in the tens place represents 90.
3. Explain the value of the digit 2 in the number 832 (it's in the tens place). The digit 2 in the tens place represents 20.
4. Explain the value of the digit 7 in the number 727 (it's in the tens place). The digit 7 in the tens place represents 70.
5. Explain the value of the digit 3 in the number 349 (it's in the tens place). The digit 3 in the tens place represents 30.
6. Explain the value of the digit 1 in the number 196 (it's in the tens place). The digit 1 in the tens place represents 10.
7. Explain the value of the digit 6 in the number 642 (it's in the tens place). The digit 6 in the tens place represents 60.
8. Explain the value of the digit 8 in the number 839 (it's in the tens place). The digit 8 in the tens place represents 80.
9. Explain the value of the digit 4 in the number 467 (it's in the tens place). The digit 4 in the tens place represents 40.
10. Explain the value of the digit 0 in the number 502 (it's in the tens place). The digit 0 in the tens place represents 0 (no tens).

WORKSHEET 12

UNDERSTANDING PLACE VALUE (ANSWERS)

Name: _____ Class: _____

1. Identify the digit in the hundreds place in the number 932. The digit in the hundreds place is 9.

2. Identify the digit in the hundreds place in the number 567. The digit in the hundreds place is 5.

3. Identify the digit in the hundreds place in the number 412. The digit in the hundreds place is 4.

4. Identify the digit in the hundreds place in the number 846. The digit in the hundreds place is 8.

5. Identify the digit in the hundreds place in the number 125. The digit in the hundreds place is 1.

6. Identify the digit in the hundreds place in the number 749. The digit in the hundreds place is 7.

7. Identify the digit in the hundreds place in the number 683. The digit in the hundreds place is 6.

8. Identify the digit in the hundreds place in the number 271. The digit in the hundreds place is 2.

9. Identify the digit in the hundreds place in the number 538. The digit in the hundreds place is 5.

10. Identify the digit in the hundreds place in the number 365. The digit in the hundreds place is 3.

WORKSHEET 13

UNDERSTANDING PLACE VALUE (ANSWERS)

Name: _____ Class: _____

1. Write a number where the digit 2 is in the hundreds place: One example is 267.

2. Write a number where the digit 2 is in the hundreds place: Another example is 298.

3. Write a number where the digit 2 is in the hundreds place: Yet another example is 222.

4. Write a number where the digit 2 is in the hundreds place: One more example is 253.

5. Write a number where the digit 2 is in the hundreds place: One more example is 225.

6. Write a number where the digit 2 is in the hundreds place: Another example is 267.

7. Write a number where the digit 2 is in the hundreds place: Another example is 292.

8. Write a number where the digit 2 is in the hundreds place: Yet another example is 216.

9. Write a number where the digit 2 is in the hundreds place: One more example is 234.

10. Write a number where the digit 2 is in the hundreds place: Another example is 289.

WORKSHEET 14

ADD THE FOLLOWING NUMBERS (ANSWERS)

Name: _____ Class: _____

1. **What is 8 + 6? 8 + 6 = 14**

2. **Subtract 7 from 15. 15 - 7 = 8**

3. **Add 9 and 5. 9 + 5 = 14**

4. **Subtract 4 from 11. 11 - 4 = 7**

5. **If you have 12 candies and you eat 3, how many candies do you have left? 12 - 3 = 9**

6. **What is 17 + 2? 17 + 2 = 19**

7. **Subtract 8 from 14. 14 - 8 = 6**

8. **Add 7 and 8. 7 + 8 = 15**

9. **Subtract 5 from 9. 9 - 5 = 4**

10. **If you have 10 stickers and your friend gives you 4 more, how many stickers do you have now? 10 + 4 = 14**

WORKSHEET 15

UNDERSTANDING THE CONCEPT OF REGROUPING (CARRYING AND BORROWING) IN TWO-DIGIT ADDITION AND SUBTRACTION (ANSWERS)

Name: _____ Class: _____

1 28 + 47 = 75

2 59 - 36 = 23

3 56 + 29 = 85

4 73 - 48 = 25

5 67 + 48 = 115

6 53 - 27 = 26

7 35 + 68 = 103

8 82 - 59 = 23

9 74 + 56 = 130

10 64 - 38 = 26

WORKSHEET 16

WORD PROBLEMS (ANSWERS)

Name: _____ Class: _____

1. There are 15 apples, and 8 are red. How many are not red? Answer: 7

2. You have 24 marbles. You give 9 to your friend. How many marbles do you have left? Answer: 15

3. There are 16 students in the classroom. 7 students are girls, and the rest are boys. How many boys are there? Answer: 9

4. There are 38 candies in one jar and 27 candies in another. How many candies are there in total? Answer: 65

5. Timmy had 14 toy cars. He got 9 more as a gift. How many toy cars does he have now? Answer: 23

6. There are 53 pencils, and 28 of them are blue. How many are not blue? Answer: 25

7. Sarah had 42 stickers. She gave 15 to her brother. How many stickers does she have now? Answer: 27

8. There are 65 books on the shelf, and 27 are fiction books. How many are non-fiction books? Answer: 38

9. Mary has 56 stickers, and Jane has 38 stickers. How many stickers do they have together? Answer: 94

10. There are 46 students on the playground. 18 of them are playing soccer, and the rest are playing basketball. How many are playing basketball? Answer: 28

WORKSHEET 17

ADDING AND SUBTRACTING WITHIN 20

Name: _____ Class: _____

1. $12 + 4 = 16$

2. $19 - 5 = 14$

3. $8 + 7 = 15$

4. $14 - 6 = 8$

5. $11 - 3 = 8$

6. $16 + 3 = 19$

7. $20 - 9 = 11$

8. $6 + 9 = 15$

9. $11 - 7 = 4$

10. $15 + 6 = 21$

WORKSHEET 18

UNDERSTANDING THE CONCEPT OF REGROUPING (CARRYING AND BORROWING) IN TWO-DIGIT ADDITION AND SUBTRACTION:

Name: _____ **Class:** _____

1. 37 + 58 = 95

2. 79 - 42 = 37

3. 65 + 38 = 103

4. 82 - 56 = 26

5. 79 + 46 = 125

6. 61 - 35 = 26

7. 57 + 68 = 125

8. 83 - 49 = 34

9. 78 + 57 = 135

10. 69 - 43 = 26

WORKSHEET 19

SOLVING WORD PROBLEMS INVOLVING ADDITION AND SUBTRACTION

Name: _____ Class: _____

1. There are 18 crayons, and 7 are red. How many are not red? Answer: 11

2. You have 28 marbles. You give 12 to your friend. How many marbles do you have left? Answer: 16

3. There are 26 students in the classroom. 9 students are girls, and the rest are boys. How many boys are there? Answer: 17

4. There are 54 candies in one jar and 28 candies in another. How many candies are there in total? Answer: 82

5. Emily had 16 stickers. She gave 8 to her brother. How many stickers does she have now? Answer: 8

6. There are 47 pencils, and 19 of them are blue. How many are not blue? Answer: 28

7. Mark had 63 toy cars, and he bought 28 more. How many toy cars does Mark have now? Answer: 91

8. There are 68 books on the shelf, and 32 are fiction books. How many are non-fiction books? Answer: 36

9. Amy has 53 marbles, and Jane has 39 marbles. How many marbles do they have together? Answer: 92

10. There are 57 students in the school gym. 25 students are playing basketball, and the rest are playing soccer. How many are playing soccer? Answer: 32

WORKSHEET 20

COMPARISON OF NUMBERS

Name: _____ Class: _____

1. ≥

2. ≤

3. =

4. ≥

5. ≤

6. =

7. ≥

8. ≤

9. =

10. =

ABOUT THE AUTHOR

Nisha is an educational professional with a fervor for storytelling and a background in science and math. She loves storytelling in her classrooms and loves to work on the social-emotional learning of young minds. She loves to create helpful content for learning and reading. She believes in making this world a better place by sensitizing people with better teaching-learning-knowing processes. Should you have any suggestions, write us at nishawrites18@gmail.com. Your feedback and suggestions are important to us. Happy reading.

www.ingramcontent.com/pod-product-compliance
Lightning Source LLC
Chambersburg PA
CBHW051938210526
45473CB00006B/2291